Mauro Dardo

Galileo

Scientist, Man

Copyright © 2015 Fabrizio Dardo
All rights reserved

ISBN: 1502399806
ISBN-13: 978-1502399809
First edition: August 2015

My father wrote this book in 2014 and finished it a few weeks before he passed away. This is why Galileo is not only written by Mauro Dardo, but also dedicated to his memory.

The cover design was going to be executed by my wife, Monica. Our four and a half year-old daughter, Chiara, was very keen to give her creative input and such was her enthusiasm for the project that she eventually proposed her own original design. My wife and I agreed that her grandfather would have loved it. The cover design of this book is therefore wholly credited to Chiara Dardo.

Fabrizio Dardo

*To the young students and
to the lovers of science*

*Galileo's story in sixty pages:
the man, the scientist, his life,
his friends, his enemies,
his affections, the triumph,
the defeat, the rehabilitation.*

Contents

1 The Age of Pisa 1

2 Professor at Padua 11

3 Return to Florence 23

4 The *Dialogue* and the Abjuration 37

5 Years of Solitude 51

Chronology

Index

Notes

1

The Age of Pisa

On February 15, 1564, three days before the death of the great Michelangelo, Galileo Galilei was born in Pisa, while Protestantism was spreading throughout Europe, and in Italy, the Roman Inquisition was prosecuting heretics.[1]

Galileo's father, Vincenzio, belonged to a noble, once-wealthy Florentine family of ancient origin. He was a man of deep culture and broad interests: he was well versed in classical languages and in mathematics and had the reputation of a talented musician and an accomplished lute player. His mother, Giulia Ammannati, was an austere and bitter woman. She married Vincenzio in 1562 and gave birth to seven[2] children, of whom Galileo was the firstborn. Having a lively disposition, Galileo inherited from his father a fighting and independent spirit, impatient of any authoritarian imposition, and from his mother a pungent character and an aggressive and sarcastic style.

In 1574, Vincenzio, along with his family, moved to Florence. Galileo was about ten years old and was sent to study rhetoric, grammar, and dialectics at the Camaldolese Monastery at Vallombrosa. Two years later he returned to Florence, where he continued his studies under the guidance of Camaldolese monks. It was during this period that he learned Greek, devoted himself to the reading of Latin authors, and became, thanks to the teachings of his father, a skilled player of the lute (so fine as to compete with the best musicians of his time). Vincenzio, who was aware of the talent of his eldest son, had in mind a definite plan for his future: he would become a doctor and would thus obtain fame and wealth.

And so, in September 1581, Galileo found himself enrolled at the University of Pisa, in the faculty of "liberal arts," as a student of medicine. The teaching of this discipline was still based on the writings of the ancient physicians Hippocrates and Galen and also included philosophy. Galileo, therefore, in his student days, became familiar with Aristotle's physics and cosmology.

Fascinated by mathematics

Galileo was nineteen years old and in his second university year when a decisive event happened for his future. The Grand Duke of Tuscany came to Pisa, followed by the court, for the usual holiday from Christmas to Easter. The official mathematician in his service was Ostilio Ricci, a disciple of the famous stutterer Niccolò Tartaglia (a great expert in ballistics who was known to have found the solution to third-degree algebraic equations).

Galileo attended the lectures on Euclid's geometry that Ricci imparted to the court pages. He was utterly fascinated and discovered that he had a passion for mathematics. He asked Ricci to improve his own mathematical abilities, and that was how he plunged himself into the study of the ancient Greek mathematicians, in particular of Archimedes, for whom he always had great admiration.

A first discovery

In the same year (1583), Galileo made his first discovery in physics. His first biographer, Vincenzo Viviani,[3] tells us that one day Galileo was in the Cathedral of Pisa and saw a chandelier swinging, the consecutive swings becoming smaller and smaller as the chandelier was coming to rest. It occurred to him to investigate if by chance the timing of the oscillations were equal, no matter how large their amplitude was. But he wondered how to measure the time of each oscillation. Galileo, who was a medical student, immediately thought to measure them to the beat of the pulse. So, by measuring on his own pulse, he verified that the number of beats was the same, regardless of the amplitude of the oscillations. He had discovered the isochronism (same period of oscillation) of the pendulum, a topic whose study he would deepen in subsequent years.[4]

Returned home

In 1585, Galileo finally gave up his medical studies, without completing his degree. He left Pisa and returned to his family in Florence. Without a university degree or a job, he began to give private lessons in mathematics in Florence and Siena. In the meantime, he enriched his humanistic culture, writing a comparison of the poetry of Ariosto and Tasso. He also delivered two lectures on Dante's *Inferno* to the Florentine Academy.[5]

His main interest, however, was in natural philosophy (the study of nature, that is, "physics"). In 1586 he wrote his first scientific essay, inspired by his studies of Archimedes's works and titled *La Bilancetta*. Here he described a balance to measure the specific gravity (that is, the relative density) of substances. The following year he circulated a manuscript that contained some ingenious theorems on the centers of gravity of complicated geometric solids, which earned him the esteem of the most famous mathematicians of the time. Among the admirers of the young scientist were the mathematicians Marquis Guidobaldo del Monte and Christopher Clavius (of the Jesuit Roman College). Guidobaldo recommended Galileo to his brother, Cardinal Francesco Del Monte, who, in turn, introduced him to Ferdinand I, the Grand Duke of Tuscany. Thanks to this powerful support, in 1589, Galileo obtained the long-awaited employment: a lectureship of mathematics at the University of Pisa, which he held for three years.

Lecturer at Pisa

Galileo's salary was sixty florins a year—just enough to survive and much lower than that of other colleagues (the professor of medicine, for example, earned two thousand florins a year). He taught Euclid's geometry and the astronomy of Ptolemy, who asserted that the earth was at the center of the universe and the sun, the moon, and the five planets known at the time (Mercury, Venus, Mars, Jupiter, and Saturn) revolved around it (the so-called "geocentric system"). He did this even though about forty years had elapsed since the publication of the famous work of Copernicus *De Revolutionibus*, in which he had proposed the "heliocentric system," with the sun at the center of the universe and with the earth and the planets revolving around it.

But what was Galileo's real opinion about astronomy? Was he a staunch follower of Ptolemy? Some scholars think that he was approaching the Copernican cosmology, and that this evolution was determined by his studies on the science of motion that had started in those years.

From the leaning tower

Galileo thought that the study of motion was the basis

for the description of natural phenomena. His views during the Pisan period, which were in contrast with those of Aristotle, are contained in a manuscript that bears the title *De Motu*. In it there is the rejection of Aristotle's idea that the speed with which bodies of the same material fall, when they fall through the same medium, is proportional to their weight. According to Galileo, the speed is the same for all bodies, and does not depend on their weight.

He is said to have performed a public demonstration of his idea: he simultaneously dropped two objects of different weight from the top of the Leaning Tower of Pisa, and everyone had to recognize that they reached the ground at the same time. Here was Viviani's account of the event [3]:

> To the great discomfort of all the philosophers, he showed…that very many of Aristotle's conclusions about motion were wrong…as (among others) that unequal weights of the same material, were falling through the same medium, not according to the proportion of their weights…but with the same speeds. He demonstrated this with repeated experiments, made from the top of the [Leaning Tower of Pisa], in the presence of other professors and all the students.[6]

Life in Pisa

In Pisa, Galileo did not only teach and devote himself to the study of motion. His exuberant vitality, characterized by "physical strength and sensuality…as well as a strong, vibrant, and joyful temper," kept him very busy. He continued to take an interest and passion for poetry, and with friends, "he liked to be often at banquets…and was particularly attracted by the variety of exquisite wines." [3] His pungent style inspired an attack against academic costumes: he wrote an irreverent text, titled *Against the Donning of the Gown*, which antagonized many colleagues.

But soon, important events happened for his future. In 1591 his father died, and the weight of providing for his large family fell on his shoulders. He then was in need of finding a job more rewarding. He turned back to his powerful patron, Guidobaldo del Monte, who recommended him to the University of Padua (the university of the *Serenissima* Republic of Venice), where there was a chair of mathematics position that was vacant. Galileo went to Venice in September 1592 to report to the authorities of the university. He was able to elicit such sympathy that he obtained the appointment he wanted. The contract provided for a salary of 180 florins a year; it was not much, but there was the hope of further increases. Thus, once he obtained the license from the Grand Duke of Tuscany, he left Pisa for Padua, and in December 1592, he began his lectures at the university. He was not yet twenty-nine.

* * *

Aristotle's universe

The Greek philosopher Aristotle (384–322 BC) believed that the universe was spherical, eternal, and finite in space, and at its center there was the stationary Earth. According to him, the universe was divided into two regions: the *supralunar* region and the *sublunar* region. The first region was unalterable and eternal. In it, the moon, the sun, and the five planets (Mercury, Venus, Mars, Jupiter, and Saturn) were attached to seven crystal spheres, concentric to the earth, while the fixed stars were on the eighth, outermost sphere. Everything was composed of an intangible substance, unalterable and transparent, named *ether* (or *quintessence*). In the sublunar world (below the sphere of the moon) there was the earth and the terrestrial bodies, composed of alterable substance, made up of four elements: earth, air, water, and fire.

For Aristotle there were two types of motion: the *natural* and the *violent*. The celestial spheres were kept in perfect and eternal circular motion by an omnipotent being, which was located beyond all them. In the sublunar region, however, each element had its own natural place and tended spontaneously to join it. Therefore, the fire rose toward the extreme of this region, while the earth tended toward the center of the universe (that is, the center of the terrestrial globe). The terrestrial bodies moved vertically in a straight line, but their motions were not perfect because they had a beginning and an end. They were caused by an intrinsic quality of the bodies that tried to bring them back to their natural place, whereby a heavy body fell to the earth because it had the quality of *heaviness* (also called *weight* or *gravity*), while *lightness* was the quality of light bodies, pushed upward. Also, an object that fell toward the earth had a speed that was proportional to its weight.

To remove a body from its natural place, an action of an external force was required. In this case the motion was *violent* and stopped only when the external force stopped acting. Once launched, a body continued to remain in motion because it was continuously pushed by the medium in which it moved: the medium filling the void left by the body itself during its motion. So, in Aristotle's physics, there was no room for emptiness.

Ptolemy's geocentric system

For more than fourteen centuries, astronomers accepted the classic work of Claudius Ptolemy (100–

170 AD), titled *Almagest*, as the foundation of the science of astronomy. The work was a synthesis of Greek astronomy that had developed in the famous Hellenistic school of Alexandria (in Egypt).

In the *geocentric system* of Ptolemy, the earth was a sphere, situated in the center of Aristotle's crystal spheres. Each of them moved on circular orbits: the sun, the moon, and the five planets. Each heavenly body described a circle, called an *epicycle*, the center of which revolved around the earth along a larger circle, called *deferent*. Thus, each body was seen moving in uniform circular motion around a point, located near the center of the earth, called *equant*. This complicated system best described the motion of heavenly bodies—but with some imperfections (for example, it predicted a motion of the moon that was very different from that observed). Finally, the outermost sphere was that of the fixed stars: it rotated on itself and produced the precession of the equinoxes.

A thought experiment

If there are two bodies of which one is falling more swiftly than the other, a combination of the two bodies will move more slowly than that part which by itself moved more swiftly, but the combination will move more swiftly than the part which by itself moved more slowly...But this would be self-contradictory, and inevitably leads to the conclusion that bodies of the same material have the same speed.

Galileo, *De Motu*

Galileo, the man

The red-haired Galileo "was of lively appearance and disposition...squarely built, of average stature...and a very strong man…[It] seemed to him that the city was in a way the prison for speculative minds, and that the freedom of the country was the book of nature which is always open to those who, with the eyes of the intellect, liked to read and study it.…"[3] "[He was] pugnacious rather than belligerent, he refrained from starting polemic battles but was ruthless in their prosecution when he answered an attack at all."[7]

Guidobaldo and Francesco Del Monte

Guidobaldo Del Monte (1545–1607) inherited his father's title of Marquis of Monte Baroccio. He studied mathematics at the University of Padua and then went to fight in Turkey and Hungary. On his return, he retired to his castle of Monte Baroccio, in the region of Urbino, to devote himself to mathematics, mechanics, and astronomy.

His brother, Francesco Maria (1549–1627), who became a cardinal in 1588, was a passionate supporter of the arts and sciences. He owned an important art collection (he gave support to Caravaggio in the early years of his work as a painter). In his palace in Rome, he set up a workshop, where he performed experiments in chemistry and alchemy. Together with his brother, he helped Galileo win a lectureship in Pisa and a chair of mathematics in Padua.

Against the Donning of the Gown

And pray don't think I'll ever don a gown
As if I were a Pharisaic professor:
I couldn't be convinced, not for a golden crown…

Remember that, like others, this remark
Was composed by the cunning and astute
To fool those simpletons whose hallmark

Is to hold as more learned and acute

*A don over another, if it's luscious velvet
Or simple felt he's chosen for his doctor's suit...*[8]

* * *

2

Professor at Padua

Galileo would remain in Padua for eighteen years: "The happiest years of my life," as he later described them. At that time, the Paduan university was one of the most renowned throughout Europe. It was open to new ideas, and it benefited from a wide freedom of thinking, which the enlightened government of the Venetian Republic was able to ensure to the students who attended it. They enjoyed a special immunity from the Roman Inquisition, and the foreigners could preserve their customs and their religious beliefs. The teachings were characterized by a spirit of freedom, especially the scientific ones.

Galileo's regular task at the university was to teach traditional topics, such as Euclid's geometry, Aristotle's mechanics, and Ptolemy's astronomy. Instead, in his large house, located in *Borgo de Vignali* at the *Santi*, he offered private tuitions on subjects such as military architecture, fortification, and perspective. These lectures were followed by young noblemen coming from the best families of Europe who were attracted by the fame of the Paduan university and the reputation of the young professor. (To accommodate these students, Galileo organized a small private college in his house.)

He also set up a small workshop where, under his

direction, a very experienced craftsman, Marcantonio Mazzoleni, built measuring instruments, which were then sold to students, soldiers, and artisans, as well as to the world-famous Venetian *Arsenale* for navigation. In 1597, he built an ingenious "geometric and military compass," which functioned both as a compass and as a slide rule. Its use spread rapidly (its price was sixty florins, but those who bought it had to complete a course of fifteen days that cost 250 florins). He dealt with thermal phenomena and invented an air thermometer. He was also granted a patent for a device that would rise water for irrigation. Additionally, he built magnets, some of which were of considerable dimensions (magnetic phenomena had become very fashionable, since the publication in 1600 of the book *De Magnete* by the Englishman William Gilbert).

Life in Padua

At Padua, Galileo's activities were challenging: lessons at the university, tuitions in his house, experiments and research in his laboratory, and meetings and discussions with the learned people of the city and with colleagues. In particular, he met with Giovanni Vincenzo Pinelli (owner of one of the finest private libraries in Europe) and Cesare Cremonini, the defender of the Paduan Aristotelian tradition.

No less challenging was his private life. A few years after his arrival in Padua, he formed a romantic relationship with a beautiful young Venetian woman, Marina Gamba. For about ten years, she would be to him an affectionate partner. From their union, two daughters and a son were born: baptized Virginia, Livia, and Vincenzio. Thus, Galileo was to provide for the needs of two families: on the one hand, his shrewish and demanding mother, his brother musician (perpetually penniless and always looking for an accommodation), and his sisters, Livia and Virginia (for them he had to provide a very expensive dowry); on the other hand, Marina and their children. Chronically pressed for money, Galileo sought to supplement his salary as a professor with private lessons and with the sale of instruments. (He could earn as much in one month from the fees of the student boarders as he earned in several months at the university.) Despite this hyperactive life, he never lost his good mood, the joy of living, and full confidence in himself.

And...in Venice

Galileo frequently went to Venice, where he had the reputation of a *bon vivant*, appreciated for his joyful and full-of-verve temper, and for his culture that was not limited to the sciences but also embraced literature and music:

> [Having an] open and vigorous temperament, he [threw himself] with the same passion in a discussion on poetry, in a love affair...or in the study of a new natural phenomenon.[9]

He was a friend of rich Venetian aristocrats, especially of the nobleman Giovanni Francesco Sagredo, whom Galileo would make immortal in one of his two masterpieces, the *Dialogue*, and Father Paolo Sarpi, a mathematician and astronomer, who served the Venetian Republic as an adviser to theological matter.

In addition to participating in the brilliant Venetian life, he held contacts with the engineers and craftsmen of the *Arsenale*. For them, he was a consultant and a supplier of instruments. For him, however, the problems of armaments and the navigation system were sources of deep scientific reflections. He wrote in the introduction of his great book, titled *Two New Sciences*[10]:

The constant activity which you Venetians display in your famous shipyard suggests to the studious mind a large field of philosophizing, especially the part that involves mechanics. For in this department all types of instruments and machines are constantly being constructed by many artisans, among whom there must be some who…have acquired the highest expertise and the most refined reasoning ability.

A new star

During the last years of the century, despite the fact that he would continue to teach the theory of Ptolemy at the university, Galileo now joined the Copernican ideas. Thus, he expressed himself privately in a letter of 1597, sent to the German astronomer Johannes Kepler[11]:

A few years ago [I accepted], like you, the doctrine of Copernicus…I have developed studies on this topic with numerous evidence in favor, but so far, I have not dared to publish them…I would have the courage to publish my ideas if there were people who think like you; but since there are none, I prefer to wait.

In September 1604, Father Ilario Altobelli of Verona informed Galileo that he had observed in the sky, in the constellation Serpentarius (Ophiuchus), the sudden appearance of a luminous body. In October, Galileo saw the same in Padua. It was not the first time that such a phenomenon had appeared: the Danish astronomer Tycho Brahe had observed one brighter in 1572. The new body was visible to the naked eye for eighteen months, during which its luminosity, after reaching a maximum, decreased continuously.

This sudden appearance aroused great wonder and

fueled many superstitious beliefs. A crowd of students and ordinary people attended the three open lectures that Galileo devoted to this phenomenon, in which, for the first time, he expressed in public his ideas, which were in contrast with those of Aristotle. He suggested, in fact, that it could be a "new star," located far away from earth, much farther than the moon and planets. It turned out that something had changed in the heavens, which could not be, as Aristotle claimed, immutable and incorruptible.[12]

Motion

Among the researches in which Galileo was committed during the years of his stay in Padua, the most important dealt with the nature of motion. He discovered the law of free-falling bodies (chapter 5). He showed that the distances travelled in equal times. When a body fell from rest, near the surface of the earth, were as the odd numbers (1, 3, 5…), and that the total distances travelled since the start of the motion were proportional to the square of the time elapsed (1, 4, 9…) He then showed that if you did not take into account the resistance of the medium (for example, the air), the falling acceleration was the same for all bodies (he announced these discoveries in a famous letter to Paolo Sarpi on October 16, 1604).

In the same years, he studied the parabolic motion of projectiles and performed experiments with inclined planes (verifying the law of free fall) and pendulums. He also formulated an approximation to the law of inertia, which he would use for subsequent discoveries. However, these results were not the subject of any publication, but they formed the basis of the theories which would appear decades later in the *Two New Sciences*.

The telescope

During the spring of 1609, Galileo was informed that in the Netherlands a small telescope had been built. He immediately built a new one in his workshop, which magnified nine times the size of the observed objects. Others followed, with enlargements bigger and bigger.

In the month of August, from the top of the bell tower of *San Marco* in Venice, Galileo showed the *doge* and other officials from the government the use of his telescope, arousing great interest. A patrician wrote in his diary:

> I went on the bell tower of [San Marco]…to see the wonders and unique effects of the [telescope] of the said Galileo…with which each of us…distinctly saw over Lizza-Fusina and Marghera, and also Chioggia, Treviso and even Conegliano, the bell tower and domes with the facade of the church of Santa Giustina in Padua; we distinguished people who came and went from the church of San Giacomo in Murano[2].

Only a few days later, Galileo offered his telescope to the *doge* of the *Serenissima*, Leonardo Donato, with a letter in which he described its practical advantages:

> Being able to discover the ships of the enemy at sea, at a distance greater than usual, and discover, with an advance of two hours or more, the number and the quality of its vessels.[13]

The Venetian government recognized the great usefulness of the instrument, renewed Galileo's teaching contract, and granted him a salary increase from 520 (the outcome of wage negotiations in 1606) to one thousand florins a year.

The wonders of the heavens

In November, Galileo built a telescope that magnified twenty times. From that moment he turned his telescope to the heavens, "in about two months...he made more discoveries...than anyone has ever made before or since."

He decided to give notice of them with a booklet of fifty-six pages, written in Latin (at that time, it was the official language of science), titled *Sidereus Nuncius*, that is, *Starry Messenger*, which appeared in print in Venice on March 12, 1610. And here are the "highly admirable phenomena" that Galileo observed:

> Looking at the moon, he "discovered a surface uneven, full of cavities and prominences, in the shape of the earth." The moon was not an immaterial, perfect substance (the *quintessence* of Aristotle), but it was a material body, just like the earth, with mountains, valleys, and craters.

> He watched a lot of stars too faint to be seen to the naked eye in the constellation of the Pleiades: he noticed thirty-six stars while, to the naked eye, people saw only six.

> Turning the telescope toward the Milky Way, he observed that it was composed entirely of "innumerable stars planted together in clusters." He then concluded that stars were much more distant and numerous than had been previously believed.

During the night of January 7, 1610, he looked at Jupiter with a telescope that magnified thirty times the size of the objects, and saw that:

> Three little stars, small but very bright, were near the planet. Although I believed them to belong to the number of the fixed stars, yet they made me wonder somewhat, because

> they seemed to be arranged exactly in a straight line parallel to the ecliptic, and to be brighter than the rest of the stars. On the east side there were two stars, and a single one toward the west.[10]

On the night of January 13, he watched four stars of Jupiter, and thus he concluded that there were:

> Four stars...circling around Jupiter (like the moon around the earth) while the whole system travels over a mighty orbit around the sun in the period of twelve years. [10]

These four luminous bodies were the largest satellites that revolved around Jupiter: Io, Europa, Ganymede, and Callisto. This was a discovery of the greatest importance, because it proved that the earth was not the only center of the celestial movements, such as the followers of Aristotle claimed.

Meanwhile, Galileo continued his astronomical observations. On the night of July 25 he noticed that the planet Saturn appeared composed of three stars. He communicated this observation to the grand duke in Florence and forwarded an anagram to Kepler, which when translated read: "I have observed that the highest of the planets is trigeminal" (the two bodies that accompanied Saturn, placed one on the east and the other on the west, seemed to touch it[14]). During the months of July and August, he watched the sun and realized that this celestial body, which the Aristotelian physics considered as a symbol of perfection of the heavens, was instead sprinkled of spots.

The Medicean stars

Galileo realized that teaching at the university, with the private lessons and other activities, took away too much time from his astronomical observations and researches. On May 7, 1610, he wrote to the secretary of state of the Grand Duke of Tuscany: "The private lessons and the pupils are a hindrance and delay to my studies, and I would like to live totally free from them."[13]

He could not, however, hope that the Venetian government would provide an adequate salary only for his scientific studies. He then entered into contact with one of his old students, the Grand Duke Cosimo II de' Medici, who had succeeded his father Ferdinand in February 1609. His aim was to get an accommodation in Florence. He then decided to dedicate to the Medici family the discovery of Jupiter's satellites, calling them "Medicean stars." He also sent to Cosimo a copy of his *Starry Messenger* and a telescope. The grand duke was very grateful for the homage and sent to Galileo, as a gift, a gold chain and a medal. On July 10, 1610, he named him "First Mathematician of the University of Pisa and First Philosopher of His Serene Highness," without the obligation to teach and reside in Pisa. The annual salary was one thousand florins of Florence (50 percent more than the one thousand Venetian florins). And so, finally, Galileo had found his patron!

The decision to leave the University of Padua angered the senate of the Venetian Republic and worried Galileo's friends. In a letter, Sagredo wrote:

> Where you can find a freedom…as in Venice?…[Here]…you did not have to serve than yourself, almost like a monarch of the universe…Who can promise you that in the stormy sea of the Court you will not be overwhelmed by the agitated winds of the emulation?[2]

But Galileo had already decided. He left Padua and arrived in Florence on September 12. He had parted amicably from Marina, leaving her in custody for some time of their son, Vincenzio. (Marina would die two years later, and Vincenzio would be entrusted to a Paduan family.) His oldest daughter, Virginia, was already in Florence with her grandmother, Giulia; only Livia accompanied him.

* * *

The heliocentric system of Copernicus

In 1543, the famous book of the Polish astronomer Nicolaus Copernicus (1473–1543), titled *De Revolutionibus Orbium Coelestium* (*On the Revolution of the Celestial Spheres*), saw the light of day. The work was divided into six parts. In the first part, Copernicus gave a general idea of his new cosmological system; the second part developed the spherical astronomy; the third, the precession of the equinoxes and the earth's motion; the fourth presented the theory of the motion of the moon (around the earth); the fifth and the sixth, the motion of the planets.

Copernicus stated first that the world had the shape of a sphere, whose center was located in the sun. The earth, which was also spherical in shape, rotated daily on its axis and annually moved on a circular orbit around the sun. Using the observations at his disposal, Copernicus was able to determine the average distances of the planets from the sun, obtaining values very close to the modern ones. He concluded that the orbits of the planets Mercury and Venus were smaller than the orbit of the earth; while Mars, Jupiter, and Saturn moved in orbits outside that of Earth. But, unlike the earth, which runs along a circular orbit around the sun, for each of the five planets Copernicus needed to add an *epicycle* to account for small variations in the speed along their orbits. Therefore, each planet moved on an *epicycle*, whose center described a larger circle around the sun.

Galileo's astrology

In Padua, Galileo also taught astronomy to medical students, since doctors needed to cast horoscopes in order to see what the stars foretold for their patients, as an aid to diagnosis and treatment. So he familiarized himself with astrology, becoming "…famous for his great ability in the art, so that distinguished people consulted him with complete confidence, in many cases asking for horoscopes and predictions."[15]

Although he had not much sympathy for astrology, he prepared many horoscopes, including those for his daughter Virginia and his friends Sagredo, Cosimo de' Medici, and Grand Duke Ferdinand I. (In January 1609 the grand duke lay ill. On the request of Grand Duchess Christina, Galileo prepared a horoscope that forecasted many more happy years for her husband. Unfortunately, the grand duke died just three weeks later.)

Sagredo

Giovanni Francesco Sagredo (1571–1620) was a Venetian nobleman and a talented amateur scientist. He studied with Galileo in Padua and was also a correspondent of the English scientist William Gilbert. He was a man at ease with the world, well able to enjoy life; he lived in Venice, in a magnificent palace on the *Canal Grande*. Sagredo wrote of himself:

> I spend my time in serving God and my country, and being free from family duties, I spend much of it in conversation, service and satisfaction of friends.[2]

Sagredo, Galileo, and many powerful and educated friends (among others: Father Paolo Sarpi and Niccolò Contarini, the superintendent of the *Arsenale*) attended academies and intellectual circles.

At that time, the most coveted circle was called "*Ridotto Morosini*," because it was founded by Andrea Morosini (a historian of the Venetian Republic) and was housed in his palace on the *Canal Grande* at *San Luca*.

Weekends in Venice

Many times, during the weekend, Galileo plunged himself into the brilliant and vibrant atmosphere of the cosmopolitan city of Venice. Here he gave vent to his passions: poetry, art, music, sprees with friends, and beautiful women.

> In addition to Sagredo, he meets regularly two jokers as Girolamo Magagnati and Traiano Boccalini. They eat and drink together, both in town and on the island of Murano, where the first produces precious glasses...from a "casino on the Canal Grande" they are used to drink "cups very full of good and sparkling cold liquor, even for those poor boatmen going up and down, toiling and sweating on their boats."[2]

* * *

3

Return to Florence

Soon after his arrival in Florence, Galileo resumed his astronomical observations. During the months of November and December, he observed Venus and recognized that it showed phases just like the moon. To protect the priority of his discovery, Galileo announced it by sending another anagram to Giuliano de' Medici (he was the brother of Grand Duke Cosimo and the Tuscan ambassador in Prague), who immediately forwarded it to Kepler.[16]

However, there were differences between the phases of Venus and those of the moon. While the apparent size of the lunar disc kept almost unchanged, the disk of Venus seemed small when it was full and larger when it was in the shape of a crescent. This meant that Venus revolved around the sun, whereas the moon revolved around the earth. In fact, when Venus appeared full, it was little beyond the sun, at the maximum distance from the earth, and when it appeared as a thinning crescent, it stood between the sun and the earth. From this discovery, and from his calculations on the satellites of Jupiter, Galileo deduced that the center of revolutions of Venus and Jupiter was not the earth but the sun, and that the same could be for the other planets (as believed by Copernicus and Kepler).

Meanwhile, the *Starry Messenger* had aroused a great interest (the 550 copies of the first edition were sold in a week), and Galileo's fame spread to distant countries. A Polish gentleman wrote to him: "Our age will be celebrated throughout the world as that of antiquity. I…rejoice to think that your name is destined to immortality, and that will be honored and admired by all."[17]

At the same time, the publication of the little book also aroused criticism. Galileo's colleague and friend from Padua, Cesare Cremonini, was determined to support the Aristotelian theory of the celestial spheres and declared that it was not worth considering what could be seen through a telescope. The German Jesuit Christopher Clavius (the leading mathematician of the Roman College and an estimator of Galileo since the age of Pisa, chapter 1) argued that Galileo's discoveries were nothing but optical illusions due to imperfections in the lenses of the telescope. However, after making careful observations, he was forced to admit that Galileo was right. Kepler himself was initially skeptical. But after observing the satellites of Jupiter with a telescope that Galileo himself had sent as a gift to the elector of Cologne (and, with an astute flair, to other European sovereigns), he wrote a pamphlet, wherein appeared the famous phrase: "*Vicisti, Galilaee!*" ("You won, Galileo!")

Triumphant welcome

The mathematicians of the Roman College (the educational institution created in 1551 by Ignatius of Loyola, the founder of the Society of Jesus) were the highest scientific authorities of the time. Galileo, therefore, was fully aware of the importance to obtain their recognition of his astronomical discoveries as well as that of the highest authorities of the Roman Church. He then asked the grand duke's permission to take a trip to Rome. He left Florence in the second half of March 1611 and arrived in Rome on Holy Thursday, the twenty-sixth of the same month.

He remained in Rome about two months as a guest of the ambassador of the Grand Duke of Tuscany. Several cardinals and he same Pope Paul V received him with great courtesy and kindness (he dispensed him to remain kneeling at the audience, as the etiquette required). In the gardens of the *Palazzo del Quirinale* (the residence of the pope), cardinals, bishops, and prelates rushed to observe the extraordinary novelties of the heavens through his telescope. Prince Federico Cesi, an influential exponent of the Roman scientific world, appointed Galileo as a member of the Academy of the Lynxes (*Accademia dei Lincei*, founded in 1603 by Cesi himself). On April 14, Galileo attended a banquet given by the academy in his honor. In May, the Jesuit fathers organized a public conference at the Roman College. In the presence of Galileo himself, three cardinals, and many prelates, they officially confirmed the validity of his astronomical discoveries (they too had observed with a telescope the Jupiter's satellites and the phases of Venus).

However, they were reticent about their interpretation. They refused to see in them a proof of the theory of Copernicus. The most authoritative representative of the spirit of the Counter-Reformation, the Jesuit Cardinal Roberto Bellarmino, was suspicious and worried. He wondered whether Galileo's discoveries could raise difficulties for theology. Galileo returned to Florence in June of 1611, proud to have received a triumphant welcome in Rome. This was, however, a triumph under surveillance!

Sunspots

Galileo lost no opportunity to embark himself on new controversies. He had, as always, an excessively good opinion of himself and was driven by the desire to assert his superiority. The first controversy arose from a discussion that took place in the palace of the grand duke. It concerned the buoyancy of bodies. Galileo wrote an essay, where he explained the phenomenon with clear demonstrations, based on the Archimedes principle of hydrostatics. For the first time, he explicitly stated in writing his opposition to the physics of Aristotle.

Shortly afterward, another controversy arose and it was much more challenging. It was about the sunspots that Galileo had stated to friends that he had seen, for the first time, a few days before his return to Florence (chapter 2). He claimed to have been the first to observe them through a telescope. There arose a bitter and nasty controversy with a German Jesuit, Father Christoph Scheiner, a mathematician and astronomer at the University of Ingolstadt. Scheiner claimed the priority of the discovery and interpreted the spots as if they were shadows of little planets revolving around the sun (and the Jesuits of the Roman College sided with him).

In 1613, the Academy of the Lynxes published three letters in which Galileo persisted to support having been the first to observe the spots and rejected the hypothesis that they were the shadows cast by the sun on some kind of planets. According to him, the spots were located on the surface of the star and were an integral part of it. The sun, therefore, was subject to changes—another argument against the Aristotelian doctrine about the incorruptible and unchangeable heavens. In addition, Galileo argued that the movements of the spots across the surface of the sun were proof that the star rotated on its axis and that the earth possessed a motion of revolution around it. To Galileo, everything felt that the Copernican system was more valid than that of Ptolemy. By now, it was clear to everyone that Galileo had become an opponent of Aristotle and Ptolemy and a staunch supporter of Copernicus.

Family problems

In the same year, 1613, Galileo separated from his two daughters, Virginia and Livia. After his transfer from Padua to Florence, he had thought of leaving them to his mother, Giulia. Soon he had to abandon this idea because of the difficult character of the old woman. To take them with him was not socially suitable: they were illegitimate and he never wanted to recognize them. But, above all, he wanted to be free from family commitments and to devote himself completely to his studies. He then thought of putting them in a convent. So it was in the autumn of the same year, Virginia and Livia entered the convent of *San Matteo* in Arcetri, just outside Florence. At the age of sixteen they took the veil: Virginia under the name of Sister Maria Celeste and Livia under that of Sister Arcangela. Virginia would adapt with resignation to the monastic world. Livia led an unhappy and melancholy life; she would never forgive the imposition of her father. ("The greatest men, often, have some character trait which is not worthy of their greatness." [17])

Denounced to the holy office

Free from the commitments of teaching and cares about his family, Galileo devoted himself to an ambitious project: to get the support of the Catholic Church for the doctrine of Copernicus. In two famous letters, one dated 1613 and directed to one of his disciples, the Benedictine friar Benedetto Castelli, and the other in 1615 and addressed to Grand Duchess Christina of Loraine (the mother of Grand Duke Cosimo II), he tried to reconcile the Copernicanism with the doctrines of the Church.[18] The two letters aroused a great stir among the academicians who opposed the Copernican system and scandalized the ecclesiastics of Florence. A Dominican preacher, who had attacked the Copernicanism from the pulpit of Santa Maria Novella, denounced Galileo to the Inquisition.

He then consulted some prelates of the Roman Curia, who suggested prudence and advised him to present the theory of Copernicus as a mere hypothesis, without interfering in the interpretation of scripture. Cardinal Maffeo Barberini, who had always showed great admiration and sympathy for Galileo, suggested him to "speak cautiously, as a mathematician"; and another friend, Monsignor Piero Dini, wrote to him kindly: "You can write freely until you stay out of the sacristy." But Galileo replied that he did not want to reduce the theory of Copernicus to pure speculation, without any scientific validity. And the more combative than ever, convinced of his thesis and sure of his oratorical skills, he decided to go personally to Rome to defend his cause.

Many friends tried to dissuade him. The Florentine ambassador in Rome, Piero Guicciardini, wrote to a minister of the grand duke in order to discourage the journey:

> "In times like these, this is not a country where you can come to discuss about the moon, or to defend or introduce new ideas."[2]

The same Cardinal Bellarmino, eager to avoid a public conflict, intervened in the same direction. But there was no way. Galileo was anxious to go to defend his case.

Copernicus on the Index

Galileo, with a personal letter from Grand Duke Cosimo for the pope in his pocket, arrived in Rome on December 3, 1615 (he had to postpone the trip for a few months, due to a strong attack of arthritis that forced him to bed in Florence). He stayed at the embassy of Tuscany, the magnificent *Villa Medici*, near *Trinità dei Monti*. (The grand duke had ordered the ambassador to receive him in an apartment "worthy and comfortable…and to make available to him a secretary, a valet, and a mule."[2])

Upon his arriving in Rome, he waged a campaign of persuasion with great passion, facing his skilled foes with an abundance of arguments. He exhibited for the first time to Cardinal Maffeo Barberini his theory of the tides, which he said were caused by the motions of the earth. But after a few weeks, he realized that, despite the care with which people were polite and friendly with him, his campaign had not reached any result. He then thought to go directly to the pope. Cardinals Barberini and Del Monte tried to dissuade him, but it was a waste of time. Galileo then turned to the young Cardinal Alessandro Orsini, presenting him with the letter of recommendation of the grand duke. Orsini agreed to act as intermediary with the pope.

Paul V was an old Roman aristocrat (he was born into the Borghese family), who did not like either new ideas or intellectual subtleties. He replied to Cardinal Orsini that "it would have been good to persuade Galileo to abandon his views on [Copernicanism]" [2]. Soon afterward, the pope summoned Bellarmino and ordered him to submit the Copernican theory to the judgment of the holy office.

On February 19, 1616, the holy office subjected to a panel of eleven theologians the following two propositions: the sun is at the center of the world and immobile and the earth is neither the center of the world nor motionless but rotates on itself. Four days later, the theologians declared unanimously that the first proposition was philosophically "foolish and absurd," formally "heretical," and contrary to scripture. The second proposition deserved, from the point of view of philosophy, the same censure as the first, and theologically it was at least "erroneous" in faith.

The sentence of the Inquisition was sent to the Congregation of the Index. They published a decree as follows: Copernicus's *De Revolutionibus* had to be *suspended* until it was corrected. It could be corrected in 1620, and the uncorrected copy would remain on the *List of Prohibited Books* (*Index Librorum Prohibitorum*) for more than two hundred years. All the other books that taught the same doctrine had to be *prohibited*. (In 1619, Kepler would deplore all this in a letter: "Some people, with their reckless behavior, have come to the conclusion that the reading of Copernicus's work, which was absolutely free for eighty years, it is now prohibited, until it is corrected."[9]) The works of Galileo were not mentioned. His friends, cardinals, did not want to offend the great scientist who was protected by the Grand Duke of Tuscany's powerful and Catholic family.

The injunction

At that time, Cardinal Bellarmino was also a consultor to the holy office. Urged by the pope, on February 26 he summoned Galileo in his palace, and in the presence of the Dominican Michelangelo Seghizzi, commissary of the Inquisition, warned him to abandon the Copernican position. Soon afterward, Father Seghizzi forbade him to hold, defend, or teach those opinions by word or in writing. If he had refused to obey, he would be imprisoned. Galileo abandoned the confident and swaggering way with which he had supported the Copernican theory in public debates and in private conversations, surrendered almost without resistance, and promised to obey.

At once, rumors circulated that during his encounter with the cardinal, Galileo had secretly recanted his theories. On May 26, at his request, Cardinal Bellarmino released a statement in which he confirmed that he had advised Galileo that Copernicus's theory could not be held or defended. Moreover, the same statement belied any inference about an alleged abjuration. Galileo, thanks to this statement, departed from Rome, convinced that his good name was intact. He took with him two letters from Cardinals Orsini and Del Monte for the grand duke, in which they declared that the scientist, during his stay in Rome had won the "highest esteem" by the Sacred College.

Years of silence

In June 1616, after the danger passed, Galileo returned to Florence and the following year he moved to *Villa Segni* in Bellosguardo, on the hills of Florence. What was he to do now? He took refuge in astronomy and performed precise measurements of the periods of the Jupiter's satellites. He thought of using them for the determination of longitude, at that time a critical issue for navigation. He came into contact with the Spanish government and proposed to them his method, but they judged it unenforceable.

In 1619, he legitimized his son Vincenzio, who came to live with him. His mother, Giulia, died the following year: this was a relief for him. In August he received a lyric poem in Latin titled "*Adulatio Perniciosa*" ("*Dangerous Adulation*"), written by Cardinal Barberini, who praised his astronomical discoveries.

Galileo interpreted this poem as a sign that he was not completely fallen out of favor with the ecclesiastical hierarchy. Then came the death of Cosimo II (in 1621), which deprived him of a powerful protector.

The Assayer

Only a few years later, Galileo was again involved in a bitter controversy with a Jesuit of the Roman College: Orazio Grassi. Taking his cue from three comets that had appeared in the sky in 1618, Grassi had written a book in which he argued that the comets were celestial objects that moved like planets. Galileo responded with another book, titled *The Assayer*. In it, he presented his theory that comets would be appearances, due to sunlight.

Galileo's theory was wrong, whereas Grassi's theory was more nearly correct. However, the controversy was an opportunity for Galileo to display his new scientific method. According to him, observation, experiment, and mathematics were the solid foundations of science. Here is his famous pronounncement[19]:

> [Natural] philosophy is written in this very great book which always lies open before our eyes (I mean the universe), but one cannot understand it unless one first learns to understand the language and recognize the characters in which it is written. It is written in mathematical language and the characters are triangles, circles, and other geometric figures.

Sticking another member of the Roman College, Galileo alienated the support of the Jesuits. Father Christoph Grienberger, a pupil of Clavius, wrote years later: "If Galileo had not angered the fathers of [the Roman] College…he could continue, until the end of his days, to write what he wanted on the motion of the earth." [17]

* * *

Tycho Brahe

On November 11, 1572, the Danish astronomer Tycho Brahe (1546–1601) observed in the constellation Cassiopeia a luminous body, which in a few months surpassed Venus in brightness and was visible even in daylight. Because its position did not change, Tycho concluded that it was a star. This "new star" shone in the sky for about two years until it weakened and finally disappeared.

The King of Denmark offered as a gift to Tycho the island of Hveen, near Copenhagen, and gave him considerable means. Tycho built two observatories where he performed astronomical observations with a precision that was never possible before. In twenty years of intense activity, he accumulated a considerable amount of data on the positions of stars, planets, and comets. Among other things, he was able to study a comet that appeared in 1577, and the observations did not detect any change of its position with respect to the fixed stars. Tycho concluded that comets were very distant from the earth, and did not belong to the sublunar world, where Aristotle had placed them. Tycho also developed a cosmological system—a compromise between the geocentric and the heliocentric models. According to him, the sun and the moon and the stars revolved around a motionless earth at the center of the universe, while the other five planets revolved around the sun. Galileo did not accept Tycho's system; he considered that of Copernicus to be the right one. In addition, he did not believe that comets were celestial bodies, as Tycho and the Jesuit Grassi thought (rightly).

Kepler

The German astronomer and mathematician Johannes Kepler (1571–1630) was a staunch supporter of the Copernican system. In 1600 he moved to Benatek, near Prague, to become an assistant to Tycho Brahe, who assigned him the difficult task of calculating the orbit of Mars using data obtained from his accurate observations. From these data, Kepler discovered three empirical laws concerning planetary motions that revolutionized astronomy, revealing that the Copernican system could be described mathematically. The first two laws were published in a book titled *Astronomia Nova* (*A New Astronomy*, 1609). The first says that the orbits of the planets are ellipses with the sun at one focus; the second says that the speed of a planet increases as it approaches the sun. The third law was published in the book *Harmonices Mundi* (*The Harmony of the World*, 1619) and it is as follows: the square of the ratio between the periods of revolution of two planets is equal to the cube of the ratio of their respective average distances from the sun. Kepler also wrote a treatise, titled *De Stella Nova*, on the appearance of the new star in 1604, which was also observed by Galileo (chapter 2).

It seems that Galileo ignored Kepler's discoveries. He did not believe that the orbits of the planets were ellipses; he was still a prisoner to the metaphysical principle of circular orbits. Galileo also found it impossible to accept Kepler's hypothesis that the tides were triggered by the attraction of the moon. He thought that the tides occurred because of the earth's daily rotation around its axis and its yearly revolution around the sun. His (mistaken) theory also provided an argument in favor of the Copernican system. (It wasn't until Isaac Newton published his law of universal gravitation in 1687 that Kepler's hypothesis concerning the lunar attraction gained a firm scientific footing.)

Cardinal Bellarmino

The Jesuit Roberto Bellarmino (1542–1621) was made a cardinal in 1599. He was the leading theologian of the Counter-Reformation and a professor of controversial issues at the Roman College, as well as a consultor to the holy office. He was one of the cardinals who sentenced Giordano Bruno to death at the stake in 1600 for heresy. He had an outstanding intellectual and moral stature, living an ascetic and frugal life. Bellarmino's opinion about the Copernican theory was as follows:

> When there should be a true demonstration that the sun is at the center of the world and the earth is in the third heaven, and that the sun does not circle the earth, but the earth circle the sun, then we should go with careful consideration in explaining the scriptures which appear contrary, and say that we do not understand them, rather than saying that it is false what it is demonstrated. But I do not think that there is any such demonstration.[20]

(With these words Bellarmino answered to a Carmelite friar who supported the Copernican system.)

Galileo's stay in Rome (1616)

Ambassador Guicciardini was beginning to worry about Galileo's ruffled activism. So he wrote to the grand duke:

> Galileo has against the monks and other people who hate and persecute him...; not only he can find himself in a serious trouble but he can also make other people sink...Engaging without a valid reason the Serene House in such complications and risks is something from which You can draw...a lot of trouble...[Galileo] does not see and cannot hear what [this dispute] entails, so that it will fall into some snare and will be in danger, and with him those who support

him…This is not a trivial matter, because this man is below our protection and our responsibility. [11]

Guicciardini also complained about the prolongation of Galileo's stay in Rome, for the huge expenses endured, given his "crazy life." [2]

Dangerous Adulation

When the Moon shines and displays
Its golden procession and its gleaning fires
In its serene orbit
A strange pleasure draws us and rivets our gaze…
Or another marvels at either the heart of the Scorpio
Or the torch of the Dog Star
Or the satellites of Jupiter
Or the ears of father Saturn
Discovered by your glass, O learned Galileo.

Cardinal Maffeo Barberini[21]

* * *

4

The *Dialogue* and the Abjuration

Pope Urban VIII

In August 1623 an unexpected event revived new hopes in Galileo: the election to the papacy of Cardinal Maffeo Barberini, who took the name of Urban VIII. Galileo immediately sent a letter of congratulations to the new pope's nephew, Francesco (sending it directly to the pope would appear as a lack of respect). In it he wrote that he was happy to see the "rebirth of hope the return of culture exiled to distant borders." And a prelate of the Roman Curia wrote to him that the pope "never has ceased to have the affection [sic] he had for you in the past."

So Galileo decided to go to kiss the foot of the new pope, hoping in his heart to obtain, if not the withdrawal, at least an attenuation of the decree of 1616. He arrived in Rome on April 23, 1624. He was warmly welcomed by his Roman friends and, during his stay, he was granted six audiences with the pope, who showed him his warm friendship. He gave him gold medals and promised him a pension for his son Vincenzio. Nevertheless, he gave him no permission to ignore the decree of 1616. And as for the Copernican theory, the pope explained to him the following argument (the so-called "argument of

Urban VIII"): "Although many facts seem to prove that the earth revolves around the sun, it is possible that God, in His infinite power, has achieved the same effect by rotating the sun around the earth, as the Scriptures say." And a cardinal told him confidentially [17]:

> The Church has not condemned [Copernicanism] and will not condemn it as heretical, but only as rash; on the other hand, there is no cause to fear that anyone could ever prove it true.

From these ambiguous openings, Galileo drew the belief that he could finally reopen the discussion on cosmological theories, provided that the Copernican system would be presented as a hypothesis. Toward the middle of June he returned to Florence, eager to resume his pen and write a book about the system of the world.

The "imprimatur"

Galileo immediately began working, and with many interruptions, mainly because of recurring illnesses (he was more than sixty years old and his forces were in rapid decline), in February 1630 the book was completed. It only remained to get the *"imprimatur"* ("let it be printed") from the Roman Church. Toward the end of March, he went to Rome to put the manuscript under examination of the master of the sacred palace, the Dominican Niccolò Riccardi, who assured him that the work could get permission—after he made some changes.

Galileo returned to Florence full of hope, but soon afterward, Father Riccardi wanted to see the text again for a second review. Galileo, now suspicious, proposed that the new examination was made by the inquisitor of Florence. The proposal was accepted, provided that he would mention the famous "argument of Urban VIII" concerning the divine omnipotence. Galileo accepted Riccardi's demand and, avoiding the Roman censors, on February 21, 1632, the book, dedicated to Ferdinand II de' Medici, appeared in Florence under the title *Dialogue Concerning the Two Chief World Systems, Ptolemaic and Copernican*, universally known as the *Dialogue*.[22]

The Dialogue

Galileo wrote the five hundred pages of the *Dialogue* in vernacular Italian rather than Latin so that the book could be read not only by the academicians but also by other educated people. The dialogue takes place during four days in the picturesque setting of the Renaissance palace of his friend Sagredo, on the *Canal Grande* in Venice.

The participants are three: Salviati, Sagredo, and Simplicio. The first two (now deceased) had been friends of Galileo. Filippo Salviati was a Florentine nobleman who had hosted the famous scientist for long periods of study in his *Villa delle Selve* at Signa near Florence. In the *Dialogue*, he is the central figure, behind which lurks the same Galileo, and he embodies the new science and the Copernican ideas. Sagredo, the friend of the unforgettable Paduan years, represents an open-minded and unprejudiced man, educated and liberal, interested in the new doctrines. Simplicio, finally, is a character of pure imagination, a staunch defender of the Aristotelian doctrine; he is a sincere man but with limited views. In the *Dialogue*, Galileo, with his unrivalled Italian prose, presents the objections of his opponents through Simplicio, trying to demolish them with scientific arguments and with a biting sarcasm.

First day. On the first day the three interlocutors discuss the worldview of Aristotle, refuting the distinction between corruptible and incorruptible bodies. The new stars and the sunspots show that the heavenly bodies are either alterable or corruptible. On the other hand, the mountains on the moon show that the physical constitution of our satellite is similar to that of the earth. The existence of the satellites of Jupiter and the phases of Venus and Mercury show how the Copernican hypothesis is simpler than that of Ptolemy. Salviati (that is, Galileo) states that the movements of celestial bodies are only the circular ones: "I therefore conclude that the only circular motion can of course agree to natural bodies that make up the universe." (This implied the refusal of Kepler's laws on elliptical orbits and on the motion of the planets.[23])

Second day. The second day is dedicated to the diurnal motion of the earth around its axis. Here Galileo, in order to reply to the objections of the Aristotelians, formulates two new concepts, which will then be the basis for the modern dynamics: the concepts of inertia and relativity.

He exposes the concept of inertia as follows: an object when descending planes sloping downward moves with an acceleration, while on planes sloping upward there is a retardation. If instead the object is set in motion along an unlimited (frictionless) horizontal plane, it continues to move along a straight line at a constant speed.

The physics of Aristotle stated that a body would move only if a force was applied to it. With the concept of inertia (the tendency of a body to maintain its state of rest or motion), Galileo demolished that Aristotelian principle, demonstrating that a force is not necessary for a body to be in motion, but is only required to cause a change in its state of motion (that is, in modern language, to cause an acceleration).

The Aristotelians argued that mechanical phenomena on earth's surface occurred as if the earth was motionless: birds in flight are not far behind with respect to the earth, cannon shots toward the west are not longer than to the east, the fall of heavy bodies occurs according to the vertical and not obliquely, and so on. To these critics, Galileo replies proposing what we call the "principle of Galilean relativity" (which includes the concept of inertia). He exposes it in a famous passage of the second day.

We can thus summarize it: mechanical phenomena in the interior of a system of reference (in Galileo's passage it is a vessel) occur in the same way, either the system (the vessel) is stationary, or it moves with uniform rectilinear motion (constant velocity). Thus, the phenomena that occur on the earth's surface look the same whether the earth is still or it turns on itself. So the Aristotelians cannot consider those phenomena as an evidence of earth's motionlessness.

Third Day. The day starts with a digression on the new star of 1604, and then moves on to the annual motion of the earth. Galileo's astronomical observations—the phases of Venus and Mercury, the satellites of Jupiter, the sunspots—are all topics that show that the five planets (Mercury, Venus, Mars, Jupiter, Saturn) do not revolve around the earth but around the sun. However, this does not yet demonstrate that the earth revolves around the sun.

These arguments in favor of the Copernican system are not decisive; that is, they are not a "true demonstration" about the motions of the earth (as Cardinal Bellarmino thought, chapter 3). They are only important clues that, as Salviati says, "greatly favor [the heliocentric system]." At that time, it was not possible to arrive at a rigorous proof. Only about half a century later, scientists would have the first scientific demonstration; and, at the beginning of the eighteenth and in the nineteenth centuries, they would have astronomical and physical evidence of the annual and diurnal motions of the earth.[24]

Fourth day. The argument is the "ebb and flow of the sea," that is, the tides. Galileo mistakenly believed that tides were due to the diurnal and annual motions of the earth and stated that this phenomenon constituted a proof of the earth's motions (a conclusive argument in favor of Copernicanism). Kepler, however, thought that tides were due to the attraction of the moon. According to Galileo, Kepler's mysterious attraction had all the appearance of those occult qualities that the Aristotelians were talking about and that he had fought. (About half a century later, Isaac Newton showed that Galileo was wrong.)

Finally, Galileo put the conclusion of the book (in essence, the "argument of Urban VIII") in the mouth of Simplicio, who declares that, in spite of all the evidence that the new discoveries can produce, the omnipotence of God may have created the universe in a way different from the one that appears.

Last journey to Rome

The *Dialogue* was received with great favor: a thousand copies were printed, and Galileo sent a large number of them to friends in Italy and elsewhere in Europe. But toward the end of July 1632, by order of the pope, its diffusion was forbidden. And in September, Cardinal Antonio Barberini, the brother of the pope, ordered the inquisitor of Florence to give notice to Galileo to go to Rome, at the disposal of the commissary general of the holy office. What would explain this unexpected and brutal change?

The fiercest opponents of Galileo (in the forefront, the Jesuits), playing on the immense pride of Urban VIII, had convinced the pope that the two systems in the *Dialogue*, the geocentric and the heliocentric, were not presented in an impartial way. And that, in the naive and obtuse Simplicio, who presented the argument concerning the divine omnipotence in the conclusion of the book, Galileo had wanted to portray the same pope. He felt himself mocked and deceived. Once an admirer and friend, the pope turned into an implacable enemy.

To this we must add that, just during that summer, Urban VIII (who had been elected with the support of France) was going through a period of tension because of the political situation that had arisen (a group of cardinals were against him because, in the midst of the "War of the Thirty Years," he had given his support to the alliance between France, Bavaria, and the Protestant Sweden, to the detriment of the Habsburgs and Spain), not taking into account the criticism for his rampant nepotism.[25] He saw enemies everywhere who tried to overthrow him from the throne of Peter. Punishing Galileo meant also to restore his prestige and to prove to the Catholic world that he was the true defender of the spirit of the Counter-Reformation. So it was that Galileo, almost seventy years old, who suffered from arthritis, and (as stated in a certificate that he had drawn up by two doctors) was afflicted with a severe hernia, was forced to leave. And after having made a will, on January 20, 1633, on a litter made available to him by the grand duke, in the bitter cold of winter, Galileo undertook for the last time his journey to Rome.

Trial and condemnation

Galileo took twenty-five days to arrive in the Eternal City, part of which he had to spend in *Ponte a Centina* (on the border between the Grand Duchy of Tuscany and the Papal State), because of the quarantine imposed on travellers during an epidemic of plague. He arrived very tired in Rome on February 13, 1633, affectionately welcomed at *Villa Medici* by Francesco Niccolini, the new Tuscan ambassador, and his lovely wife, Caterina.

After two months of waiting, the first hearing took place on April 12. The prosecutors came directly to the point: Did Galileo violate the injunction of 1616?

In his defense, he produced a copy of the statement that Cardinal Bellarmino (who had died in 1621) had released to him in the month of May of that year (chapter 3). Such a statement contained the prohibition to "hold" and "defend" the Copernican doctrine but not to "hold, defend, or teach it, in any way whatever, orally or in writing" as, however, it appeared from the minutes (without any signature) relative to the meeting with the cardinal and Father Seghizzi that took place on February 26, 1616 (chapter 3). Moreover, Galileo argued (denying the evidence) that in the *Dialogue*, Copernicanism was not held to, neither was it defended, so the injunction of 1616 was not violated.

After this questioning, Galileo was housed comfortably in two rooms located in the palace of the holy office, and was informally advised by the commissary general to admit that "he was clearly wrong and that in his book he had gone too far." In his second hearing, on April 30, in fact he admitted that in the book there were arguments in favor of Copernicanism stronger than were intended by him, and that he could reconsider them "in the most effective way that the blessed God will enable me." He was thus granted to return to the Tuscan embassy. He then was more peaceful and confident that the conclusion of the trial would not be so dramatic. He was even allowed to go in a carriage through the gardens of Rome and to take a trip to Castel Gandolfo.

Galileo was called in again on June 21 to see, once for all, whether he had held Copernicus's heliocentrism after the injunction of 1616. The commissary urged him, under threat of torture, to tell the truth. Galileo replied that he was there "to obey" and persisted in claiming that "he had not held Copernicanism after 1616."

The following day the sentence of the holy office was communicated to Galileo: he was condemned to recant heliocentrism and to imprisonment for "vehement suspicion of heresy" (three out of the ten cardinals present did not sign the verdict, among them Cardinal Francesco Barberini). The *Dialogue*, in turn, had to be put on the *List of Prohibited Books* (it would remain there until 1822). In the hall of the convent of *Santa Maria Sopra Minerva*, before the cardinals and the prelates of the holy office (Urban VIII was not present in person, but he was there in spirit), the old scientist, after wearing the white robe of a penitent, pronounced, on his knees, the abjuration formula:

> With a sincere heart and unfeigned faith I abjure, curse, and detest the above mentioned errors and heresies and swear that in the future I will never again say or assert, orally or in writing, anything which might cause a similar suspicion about me; on the contrary, if I should come to know any heretic or anyone suspected of heresy, I will denounce him to this Holy Office. [10]

(Getting up after having pronounced the abjuration formula, Galileo would whisper, tapping one foot on the ground, "yet it moves." It has never been possible to give this news a historical basis.)

Back to Arcetri

The following day, Galileo was ordered not to go out of the Tuscan embassy. Seven days later, the pope gave him permission to leave the "prison" of *Villa Medici* to go to Siena, guest of the archbishop, Ascanio Piccolomini, his sincere friend. He arrived in Siena on July 7. In the archbishop's palace, he was housed in an apartment with "silk tapestries and a rich furniture" where he could receive the notables of the city, who came to show him their admiration and to discuss with him about science. His eldest daughter

(Sister Maria Celeste), the only member of his family that had followed with anxiety the trouble that he was going through, wrote to him with great affection: "When you were in Rome, I said to myself: 'if he were only at Siena!' Now that you are at Siena I say: 'If only he were at Arcetri!' But God's will be done."[26]

Piccolomini's friendship and the Siena environment were able to lift the state of dejection in which Galileo found himself. In early December, the holy office authorized him to establish his residence in Arcetri, in the villa "*Il Gioiello*," which he had rented in 1631 to be closer to his beloved daughter. The authorization provided, however, that he had to live isolated, under "house arrest." But his happiness to return to Arcetri was short-lived. At the beginning of 1634, Sister Maria Celeste became seriously ill, and she died on April 2, resulting for her poor father "an incurable suffering." He wrote to Geri Bocchineri (the brother-in-law of his son Vincenzio): "An immense sadness and melancholy, a deep inappetence make me hateful to myself. In short, I am constantly called by my beloved daughter."

Despite the great pain, Galileo was able, once again to find the strength to resume his studies and returned to his researches of mechanics, which he had never abandoned.

* * *

Urban VIII

Maffeo Barberini (1568–1644) was born into a family of wealthy Florentine merchants. He had been a pupil of the Jesuits in Florence and at the Roman College and had received a doctorate of law from the University of Pisa. At the age of thirty-eight, he was appointed cardinal and archbishop of Spoleto by Pope

Pius V. After the sudden death of Gregory XV in July of 1623, some cardinals began frantic negotiations for the election of his successor. The best candidate was Barberini, who was supported by France (he had been the apostolic nuncio to the court of King Henry IV in Paris). But they also needed the approval of the Spanish cardinals. On August 6, the agreement was reached, and Barberini was elected pope, under the name of Urban VIII. For Galileo, the election of his longtime friend was a godsend. He sent him, through Barberini's nephew, a letter of congratulation and dedicated to him *The Assayer*, the book that the Academy of the Lynxes was editing in those days. In the spring of 1624, Galileo rushed to Rome with the naive hope to resume his Copernican battle. He would change his mind bitterly nine years afterward.

The holy office and the Index

The Roman Inquisition (also called holy office) was established by Pope Paul III in 1542, three years before the Council of Trent was opened (the full name was Supreme Sacred Congregation of the Roman and Universal Inquisition). It was a tribunal composed of cardinals, prelates chosen from the Dominican Order, and consultants (scholars of theology and canon law), whose task was to maintain and defend the integrity of the faith, and to examine and proscribe errors and false doctrines.

In 1571 the Sacred Congregation of the Index was created, which dealt with the *List of Prohibited Books* (*Index Librorum Prohibitorum,* also simply referred to as *Index*) and with the aim of monitoring and banning heretical and immoral books. In 1965 (at the end of the Council Vatican II) Pope Paul VI reconstituted the holy office as the Congregation for the Doctrine of the Faith, and in 1966 the *Index* was formally abolished.

The Jesuits

The religious order of the Jesuits was founded by Ignatius of Loyola in 1540 and was the guardian of the culture within the church. Inside it there were eminent mathematicians and astronomers. In the case of Copernicus, early Jesuits seemed favorable to the new cosmological theory and tried to interpret it according to the orthodoxy of the Church. After the ban of the *De Revolutionibus* in 1616, the Jesuits were directed toward the system of Tycho Brahe (chapter 3), which seemed to reconcile the Aristotelian principle of the earth motionless at the center of the universe with the new astronomical discoveries.

The Dominicans

The Religious Order of Preachers (commonly called Dominicans) was founded in the early thirteenth century by the Spaniard Dominic de Guzman, with the aim of countering the medieval heretical doctrines by preaching and following a life of poverty. At the time of Galileo, the Dominicans, the masters of the holy office, were the vanguard of Catholicism more conservative and suspicious of any kind of innovation. Thus they became the staunchest opponents of Copernicus and Galileo. Their position toward Galileo, however, changed after 1616, as well as changed that of the Jesuits. In the events of 1616, the charge was left by the Dominicans (chapter 3), and Galileo found support among the Jesuits. In the process of 1633, the Jesuits were to instigate the prosecution, and Galileo was supported by the Dominicans who had authorized the printing of the *Dialogue*.

Niccolini's letter

After the second hearing, Galileo returned to the Tuscan embassy. "The poor man has came back more dead than alive," Ambassador Niccolini observed. On May 15, 1633, Niccolini wrote a letter to the secretary of state of the grand duke.

> In regard to what Your Most Illustrious Lordship tells me, namely that his Highness does not intend to pay for Signor Galilei's expenses here beyond the first month, I can reply that I am not about to discuss this matter with him while he is my guest, I would rather assume the burden myself.[27]

To Galileo from his daughter

> Just as suddenly and unexpectedly as word of your new torment reached me, Sire, so intensely did it pierce my soul with pain to hear the judgment that has finally been passed, denouncing your person as harshly as your book…I pray you not to leave me without the consolation of your letters, giving me reports of your condition, physically and especially spiritually. Though I conclude my writing here, I never cease to accompany you with my thoughts and prayers, calling on His Divine Majesty to grant you true peace and consolation.[27]

The sentence

> We say, pronounce, sentence, and declare that you, the above-mentioned Galileo, because of the things deduced in the trial and confessed by you as above, have rendered yourself according to this Holy Office vehemently suspected of heresy,

namely of having held and believed a doctrine which is false and contrary to the divine and Holy Scripture: that the sun is the center of the world and it does not move from east to west, and the earth moves and is not the center of the world, and that one may hold and defend as probable an opinion after it has been declared and defined contrary to Holy Scripture…[We] order that the book Dialogue…be prohibited by public edict. We condemn you to formal imprisonment in this Holy Office at our pleasure. As a salutary penance we impose on you to recite the seven penitential psalms once a week for the next three years. And we reserve the authority to moderate, change, or condone wholly or in part the above-mentioned penalties and penances.[10]

* * *

5

Years of Solitude

In Siena, the surveillance of Galileo was in fact nonexistent and allowed him to lead a social life like the old days. The Inquisition was informed, and they immediately ran for cover; in Arcetri a very strict control was reestablished. Galileo's friends who wanted to go to visit him had to get permission from the church authorities. He, who had so loved the life of society, brilliant conversations, and polemic discussions, was reduced to solitude. His only consolation was the frequent correspondence he exchanged with his admirers and friends.

His scientific work, however, had a wide circulation outside of Italy, in the countries that were not controlled by the Roman Inquisition. The *Dialogue* was translated into Latin and was published by the famous House of Elzivir, based in Leiden, the Netherlands. And so his thoughts spread among the elite of European scientists.

Meanwhile, in Arcetri, he resumed his pen and completed the last work of his career. It concerned two arguments of the years of Pisa and Padua: the resistance of materials and motion. Toward the end of 1634, the book was almost completed. It was time to get it published. He tried in Venice, Austria, Germany, and France. At last, with the help from Count de Noailles, a former student of Padua and now ambassador of France to the pope, he came into contact with Lodewijk Elzevir (the same who had published the Latin version of the *Dialogue*). Thus, in July 1638, the second of Galileo's masterpieces was published. It was titled *Discourses and Mathematical Demonstrations on Two New Sciences*, also known as *Two New Sciences*.[28] (Galileo judged it "superior to anything else I have published.")

Two New Sciences

As the *Dialogue* had been, the latest book was written in Italian. The conversation again takes place over four days, with the same participants: Salviati, Sagredo, and Simplicio.

During the first two days, the interlocutors discuss the resistance of materials to be broken, the free fall of bodies near the earth's surface, the oscillations of the pendulum, the vibrations of the strings, the sound, and the problem of scaling. In the third and fourth days (the most celebrated part of the book), they treat the new science of motion.

In the introduction of the third day, Galileo writes [28]:

> We bring forward a brand new science concerning a very old subject. There is perhaps nothing in nature older than motion, about which volumes have been written by philosophers; yet I find many essentials of it that are worth knowing which have not even been remarked, let alone demonstrated.

He then proposes a new science of motion that will have, as basic elements, mathematics and experimentation. Thus he concludes: "There will be opened a gateway and a road to a large and excellent science into which minds more piercing than mine shall penetrate to recesses still deeper."

The discussion between Sagredo, Salviati, and Simplicio begins with the definition of uniform rectilinear motion:

> Equal or uniform motion I understand to be that of which the parts run through by the moveable in any equal times whatever are equal to one another. [28]

(This means that the distances travelled in equal times are as the positive integers: 1, 2, 3…)

The definition of accelerated motion instead gives rise to an interesting discussion, which allows Galileo to establish that the speed of free fall of a heavy body is proportional to the time elapsed from rest (while at first, during the Paduan days, he had supposed that it was proportional to the space path).

> When I consider that a stone, falling from rest, successively acquires new increments of speed, why should I not believe that those additions are made by the simplest and most evident rule? For if we look into this attentively, we can discover nonsimpler addition and increase than that which is added on always in the same way. [And so] the definition of the motion of which we are going to treat may be put thus: I say that that motion is equably accelerated which, abandoning rest, adds on to itself equal moments of swiftness in equal times. [28]

From the proportionality between the speed and the time of fall, Galileo deduces that the distance travelled in a naturally accelerated motion is the same as in a uniform motion with a speed that is equal to the average between the initial speed and the final one. From this he gets that the distance travelled is proportional to the square of the elapsed time. (That is, during equal intervals of time, the distances are as the square of the times: 1, 4, 9…)

Galileo also describes a "thought experiment." Using pendulums of different lengths, he shows that if an object falls from the same height on planes of different inclination, it acquires, at the end of the descent, equal speed. So the fall along the vertical can be thought of as the limiting case, when the plane is vertical. This allows him to use inclined planes to verify experimentally the law of the square of the times in uniformly accelerated motion.

The fourth day is dedicated to the motion of the projectiles. Galileo uses the concept of inertia (chapter 4) and lays down another principle: that of the composition of displacements. Thanks to these two principles, he explains the motion of projectiles as the composition of a horizontal motion with a constant velocity and a vertical motion with a constant acceleration (the acceleration of gravity). Therefore, he shows that the trajectory is a parabola (a result unknown to his predecessors).

The success of the book was immediate. It was read throughout Europe. Galileo, however, was not able to get a copy until the following year. The Jesuits had purchased, in Prague, all copies to be sent to Italy. Toward the end of the eighteenth century, the great mathematician Louis Lagrange, speaking of the new science of terrestrial motion presented in the *Two New Sciences*, thus commented: "It needed an extraordinary genius to deduce the laws of natural phenomena that humanity had ever under the eyes, but their explanation was always escaped the research of the philosophers."

The events of the last years

Despite his inexorably deteriorated health (his daughter, Sister Maria Celeste, continually begged him not to drink too much wine, because it worsened his health[29]), Galileo continued his scientific work with undiminished commitment. In 1636 he revived for the States General of Holland his method of the satellites of Jupiter (chapter 3) for the determination of longitudes. As a tool for measuring time, he proposed a pendulum that he had planned. The Dutch took into consideration the proposal and also decided to send him a gold chain as a gift. But the news reached Rome, and the authorities of the holy office ordered the inquisitor of Florence to prevent any negotiation. So Galileo was forced to discontinue the contacts with Holland and to refuse the gift.

The 1638, the year of the publication of the *Two New Sciences*, was full of events. The first, painful, was the news that he had become completely blind. He wrote to his friend Elia Diodati:

> Here I am, Lord! Your dear friend and servant Galileo, has for the last month been completely and irremediably blind! Consider the affliction in which I find myself when I think that this heaven, this world, this universe, that my remarkable observations and clear demonstrations, had enlarged thousand times greater than had been known by men of past ages, for me is now reduced to the space it occupies my own person.[13]

The control of the ecclesiastic authorities subsided, and he could get friends and admirers who softened his loneliness (among these visits, it seems there was also that of the English poet John Milton), and he could still be involved in a final scientific controversy.

This controversy regarded the ashen light that the moon seemed to emit from its dark side. According to some philosophers, it was due to a kind of phosphorescence of the atmosphere of our satellite. Galileo, in a letter addressed to Prince Leopold of Tuscany, explained correctly that the faint light that could be seen to illuminate the moon was in fact due to the sunlight reflected from the earth.

In 1639, he was allowed to receive in his house, as a student, the seventeen-year-old Vincenzio Viviani, who would write the first biography of the scientist. Two years later, the mathematician Evangelista Torricelli, Castelli's pupil, came to live at Arcetri as Galileo's assistant. He would be his main scientific heir.

Alessandra, his last love

Galileo, who never married, had always been sensitive to feminine charms. In Pisa he had tasted the pleasures of life as a student and a young professor; in Padua, as a companion of Marina; in Venice, as a refined *bon vivant*. We do not know his female relationships in Florence, but it is certain that he carried on love affairs until late in life (his son Vincenzio reproached him for those loves that gave "opportunity to murmur"). And now, an affectionate friendship, purely platonic, burst into his life.

Alessandra Bocchineri Buonamici was thirty years old and had been twice a widow. By her third marriage, she had become the sister-in-law of his son Vincenzio. A beautiful and smart lady, she won the heart of the famous scientist; and, in turn, she was captured by the intelligence and charm that emanated from him. From Prato, where she resided, Alessandra wrote, in March 1641:

> Sometimes I wonder how I could do to find a way to meet you and spend a day in your company, without giving occasion for scandal and jealousy to those people that have prevented us from realizing this project.[9]

And Galileo, in his last letter of December 20, 1641, wrote:

> I received your letter with great pleasure. [It was] to me a great consolation, because it found me in bed, seriously ill. I thank you from the bottom of my heart for the benevolent feelings that you show to me. Please do not blame the brevity of this letter if not to the severity of my illness. I kiss your hands with love and with all my heart.[9]

In November 1641:

> It came upon a slow fever and palpitations of the heart. After two months of illness, that gradually consumed him.

During the night of January 8, 1642, "[Galileo] at age seventy-seven ten months and twenty days, with philosophical and Christian serenity, gave his soul to the Creator."[3]

On January 9, Galileo was buried privately in the Church of Santa Croce in Florence. In 1737, Galileo's tomb was finally built in the same church. Nearby, there are the tombs of Michelangelo and Machiavelli.

* * *

To Cardinal Francesco Barberini (*From the Inquisitor of Florence*)

I went in person suddenly, with a doctor...a confidant of mine, to recognize the status of Galileo in his villa in Arcetri [in order] to have light as if, coming to [Florence], he could [publicize], with meetings and speeches, his damned opinion about the motion of the earth. I found him totally sightless and although he hopes to recover, not being more than six months that the cataracts appeared in his eyes, the doctor, however, given his age of 75 years [stated that] his illness [is] almost incurable. In addition to this, he has a constant pain and [a strong insomnia]. He is so badly damaged, that he has the form of a corpse rather than that of a living person. His studies were interrupted for blindness and his conversation is not popular, because, being so in poor condition of health, he cannot do anything but talking about his illness with whom who sometimes go to visit him. I think that [if] the Holiness of Our Lord, in his infinite mercy toward him, wants to allow him to stay in [Florence], he would not have occasion for meetings. He is ill in such a manner that will be enough a good admonition to curb him.[30]

Florence, February 13, 1638

Viviani

Vincenzo Viviani (1622–1703) came to the villa *Il Gioiello* in October 1639 and remained near Galileo until his death. He was educated by the Jesuits in Florence and devoted himself to geometry and physics. After Galileo's death, he was appointed to major public offices by the Grand Duke Ferdinand II. In 1654 he wrote the first biography of the scientist, titled *Racconto istorico della vita di Galileo* (*A Historical Narrative of the Life of Galileo*). Afterward he edited the first edition of the works of Galileo.

Torricelli

Evangelista Torricelli (1608–1647) studied mathematics and worked with Benedetto Castelli, who recommended him to the illustrious scientist, suggesting that he employ him as his assistant.

And so, in October 1641, Torricelli moved to Arcetri, into Galileo's villa. A few months later Galileo died, and Grand Duke Ferdinand II appointed Torricelli as Galileo's successor as mathematician of the court. In addition to mathematics, Torricelli devoted himself to physics, studying the motions of bodies, fluids, and optics.

Alessandra

Galileo met Alessandra Bocchineri (1600–1649) in July 1630, in the villa of her husband, the diplomat Giovanfrancesco Buonamici, in the hills of Sovignano (her husband often sent Galileo the excellent wine of his *Vigna di Venere*). Galileo was completely fascinated by the beautiful and intelligent noblewoman and fell in love with her.

They exchanged a few letters, after which the mail was interrupted until 1641, when Alessandra, responding to a letter from Galileo, assured him that she wanted to accept his invitation but was forced to reject it for fear of giving scandal. Galileo wrote his last letter to her fifteen days before his death, renewing to her his affection and devotion.

To Giovanfrancesco Buonamici

Your Lordship has been so kind to send me the fruits of those hills loved by Bacchus. The two liqueurs are so suited to my taste that, without sharing them with others, I just want to enjoy them by myself. Meanwhile, I duly thank you for the gift.

From Galileo, February 14, 1634 [13]

Rehabilitation

In 1979 Pope John Paul II gave a speech in which he admitted that Galileo "had suffered much from ecclesiastical men and institutions." Then he formed a special commission to investigate the "Galileo Affair." Eleven years later, in 1992, the commission concluded its work. Pope John Paul II declared that "Galileo's trial was not merely an error but also an injustice."[31]

* * *

Chronology

1564—February 15: Galileo is born in Pisa.

1574—Galileo's family moves to Florence.

1581—September: Galileo enrolls at the University of Pisa as a student in medicine.

1583—According to Vincenzo Viviani, Galileo's first biographer, Galileo discovers the isochronism of the pendulum.

1585—Galileo leaves the university without a degree.

1586—Galileo invents a hydrostatic balance.

1587—Galileo finds certain properties of the centers of gravity.

1589—Mathematical lecturer at the University of Pisa. According to Vincenzo Viviani, Galileo performs the famous Leaning Tower demonstration on falling bodies.

1592—Galileo leaves Pisa and becomes professor of mathematics at the University of Padua.

1600—Galileo and his common-law wife, Marina Gamba, have a daughter named Virginia.

1601—A second daughter, Livia, is born to Galileo and Marina.

1604—Law of free-falling bodies. October: Galileo observes a new star and delivers three public lectures on it.

1606—A son named Vincenzio is born to Galileo and Marina.

1609—Fall: Galileo begins to observe the heavens with his telescope.

1610—January 13: Galileo discovers four satellites of Jupiter. March: the *Sidereus Nuncius* is published in Venice. July 10: Galileo is appointed "Philosopher and Chief Mathematician" to the Grand Duke of Tuscany. Summer: Galileo observes Saturn as "three-bodied." September: Galileo leaves Padua and moves toFlorence. Fall: Galileo observes the phases of Venus.

1611—March 29: Galileo arrives in Rome. *April 25*: Galileo is made a member of the *Academy of the Lynxes*. *May 13*: The Jesuits praise Galileo's astronomical discoveries.

1612—Controversy on floating bodies. Marina Gamba dies.

1613—Controversy with the German Jesuit astronomer Christoph Scheiner on sunspots. *Letter to Benedetto Castelli*.

1614—December 21: A Dominican friar preaches a sermon against Galileo.

1615—Galileo is denounced to the Holy Office. *Letter to the Grand Dichess Christina*. *December*: Galileo goes to Rome to defend his Copernican ideas.

1616—January: Declaration of the holy office on the Copernican theory. Cardinal Bellarmino warns Galileo not to hold or defend the Copernican theory. *March*: The Congregation of the Index suspends Copernicus's *De Revolutionibus* until corrected. *May*: Cardinal Bellarmino writes a statement to Galileo certifying that he had not been on trial or condemned by the Inquisition.

1619—The controversy with the Jesuit Orazio Grassi on comets begins.

1620—Cardinal Maffeo Barberini sends Galileo a poem in his honor, titled *Adulatio Perniciosa*. Galileo's mother dies.

1621—February: Grand Duke Cosimo II dies prematurely. *September*: Cardinal Bellarmino dies.

1623—August: Cardinal Maffeo Berberini is elected pope and takes the name of Urbano VIII. Galileo's *The Assayer* is dedicated to the new pope and is published in Rome sponsored by the Academy of the Lynxes.

1624—Galileo goes to Rome to pay homage to the new pope. *Fall*: Galileo begins working on a book that discusses the system of the world.

1630—Galileo completes work on his book.

1624—August: Prince Federico Cesi dies.

1632—February: The *Dialogue* is published in Florence. *Summer*: Pope Urban VIII prohibits the distribution of the book. *September*: Galileo is summoned to Rome to stand trial.

1633—February: Galileo arrives in Rome. *Spring*: The Inquisition trial proceedings begin; they are concluded on 21 June. *June 22*: Galileo is convicted of "vehement suspicion of heresy" and is condemned to the abjuration of heliocentrism and to imprisonment. The *Dialogue* is put on the list of prohibited books (the Index). Galileo recites the abjuration formula at the convent of Santa Maria Sopra Minerva. *June 23*: Galileo's prison sentence is commuted to house arrest at Villa Medici. *June 30:* The sentence is again commuted to house arrest in Siena, at the residence of the archbishop. *December 1*: The sentence is definitely commuted to house arrest at his villa in Arcetri near Florence.

1637 Galileo becomes completely blind.

1638—*July*: Galileo's *Two New Sciences* is published in Leiden, the Netherlands.

1639—Vincenzo Viviani begins studying with Galileo and assisting him.

1641—October: Evangelista Torricelli moves in with Galileo, serving as his research assistant.

1642—January 8: Galileo dies in Arcetri.

1835—The new edition of the Index for the first time omits Galileo's *Dialogue* from the list.

1979—Pope John Paul II begins an informal rehabilitation of Galileo that is concluded in 1992.

1992—Pope John Paul II admits that Galileo's trial was not merely an error but also an injustice.

Index

A

abjuration, 31, 37, 45, 63
Academy of the Lynxes, 25, 26, 47, 62
acceleration, 15, 40, 54
Adulatio Perniciosa (Dangerous Adulation) 31, 62
Alighieri, Dante, 3, 71
Almagest, 8
Altobelli, Ilario, 14
Arcetri, 27, 45, 46, 51, 56, 58, 59, 63, 69
Archimedes, 2, 3, 26
Ariosto, Ludovico, 3, 71
Aristotle, 2, 5-8, 11, 15, 17, 18, 26, 33, 39, 40
Arsenale, 12, 13, 21
Assayer, The, 32, 47, 62
Astrology, 21
Astronomia Nova (New Astronomy), 34, 73
astronomy, 4, 8, 9, 11, 21, 31, 34

B

Barberini, Antonio, Cardinal, 42
Barberini, Francesco Cardinal, 45, 58
Barberini, Maffeo Cardinal (Pope Urban VIII), 28, 29, 31, 36, 37, 47, 62
Bellarmino, Roberto, Cardinal, 25, 28, 31, 34, 35, 41, 44, 62
Boccalini, Traiano, 22
Bocchineri, Alessandra, 56, 59
Bocchineri, Geri, 46
Brahe, Tycho, 14, 33, 48
Bruno, Giordano, 35
Buonamici, Giovanfrancesco, 59, 60
Buonarroti, Michelangelo, 1, 30, 57, 71

C

Canal Grande, 21, 22, 39
Castelli, Benedetto, 27, 56, 60, 62
Cesi, Federico, Prince 25, 63

Christina of Loraine, Grand Duchess, 21, 27
Church, Roman, 24, 27, 38, 48, 51, 57
Clavius, Christopher, 4, 24, 32
College, Roman, 4, 24, 25, 26, 32, 35, 46
College, Sacred, 31
comet (s), 32, 33, 62
compass, military, 12
Contarini, Niccolò, 21
Copernican system (theory), 4, 27- 35, 37-44, 62
Copernicanism, 28, 29, 42, 44
Copernicus, Nicolaus, 4, 14, 20, 21, 24-30, 33, 44, 48,, 62
Cosimo II, de' Medici, 19, 21, 23, 27, 28, 31, 62
cosmology, 2, 4
Counter-Reformation, 25, 35, 43
Cremonini, Cesare, 12, 24
Curia, Roman, 28, 37

D

De motu, 5, 8
Del Monte, Francesco, Cardinal, 4, 9, 29, 31
Del Monte, Guidobaldo, 4, 6, 9
De Revolutionibus, 4, 20, 30, 48
Dialogue, 13, 37, 39, 42, 44, 45, 48, 50, 51, 52, 63, 73
Dini, Piero, 28
Diodati, Elia, 55
Dominicans, 48
Donato, Leonardo, 16

E

Earth, 4, 6, 7, 8, 15, 17, 18, 20, 23, 29, 30, 33-35, 38, 40-42, 50, 52, 56, 58, 73
motion of, 20, 32, 33, 40-42, 48, 58
Einstein Albert, 73
Elzevir, 52
Euclid, 2, 4, 11

F

fall, free, 15, 52, 53
Ferdinand I, de' Medici, 4, 21

Ferdinand II, de' Medici, 39, 59
Florence, 1, 3, 18-20, 23-29, 31, 38, 39, 42, 46, 55-58, 61, 63, 66, 72

G

Galen, 2
Galilei, Livia (daughter, Sister Arcangela), 12, 20, 27, 61
Galilei, Livia (sister), 13
Galilei, Vincenzio (father), 1, 2
Galilei, Vincenzio (son), 12, 19, 31,37, 46, 56, 61
Galilei, Virginia (daugter, Sister Maria Celeste), 12, 21, 27, 61
Galilei, Virginia (sister), 13
Gamba, Marina, 12, 61, 62
Geymonat, Ludovico, 72, 73
geocentric system (geocentrism), 4, 7, 8, 33, 42
geometry, 2, 4, 11, 58
Gilbert, William, 12, 21
Grassi, Orazio, 32, 33, 62
gravity, 3, 7, 54, 61, 73
Grienberger, Christoph, 32
Guicciardini, Piero, 28, 35, 36
Guzman, Dominic de, 48

H

Harmonices Mundi (*The Harmony of the World*), 34
heliocentric system (heliocentrism), 4, 20, 33, 41, 42, 44, 45, 63
heresy (heretics), 1, 35, 45, 49, 63
Holy Office (see also Inquisition), 27, 29-30, 35, 42, 44-50, 55, 62
Huygens, Christiaan, 72

I

Index, Congregation of the, 28, 30 47, 62, 63
inertia, 15, 40, 41, 54
Inquisition (see also Holy Office), 1, 11, 28, 30, 47, 51, 62, 63

J

Jesuits, 26, 32, 42, 48, 54, 58, 62
John Paul II, Pope, 60, 63
Jupiter, 4, 6, 17, 18, 20, 41
satellites of, 19, 23, 24, 25, 31, 36, 40 41, 55, 61

K

Kepler, Johannes, 14, 18, 23, 24, 30, 33, 34, 40, 42, 72-74
laws of, 40

L

Lagrange, Louis, 54
Leaning Tower (of Pisa),4, 5, 61, 72
List of Prohibited Books (Index), 28, 30, 45, 47, 63
Loyola, Ignatius of, 24, 48

M

Machiavelli, Niccolò, 57
Magagnati, Girolamo, 22
Mars, 4, 6, 20, 34, 41
mathematics, 1-4, 6, 9, 32, 52, 59, 61
Mazzoleni, Marcantonio, 12
mechanics, 9, 11, 14, 46
Medicean stars, 18, 19
Medici, de', Giuliano, 23
medicine, 2, 4, 61
Mercury, 4, 6, 20, 40, 41
Milky Way, 17
Milton, John, 55
Moon, 4, 6-8, 15, 17-18, 23, 28, 33-34, 36, 40, 42, 55, 56, 73
motion, 8, 20, 23, 33

N

Newton, Isaac, 34, 42, 74
Niccolini, Francesco, 43, 48, 49
Noailles, Count de, 52

O

Orsini. Alessandro, Cardinal, 29, 31

P

Padua, 6, 9-12, 14, 16, 19-21, 24, 27, 39, 51-53, 56, 61, 72
university of, 6, 9, 19, 61
Paul V, Pope, 25, 29
pendulum (s), 3, 15, 52, 54, 55, 61
physics (natural philosophy), 2, 3, 7, 18, 26, 40, 58, 59, 71
Piccolomini, Ascanio, Archbishop, 45, 46
Pinelli, Giovanni Vincenzio, 12
Pisa, 1-6, 9, 19, 24, 51, 56, 61, 72
university of, 2, 4, 19, 47, 61
plane (inclined), 15
planet (s), 4, 6, 8, 15, 17-18, 20, 23, 26, 33-34, 40, 41, 73, 74
Pleiades, 17
Prague, 23, 33, 54
projectile (s), 15, 54
Protestantism, 1
Ptolemy, 4, 7-8, 11, 14, 26, 40

Q

Quintessence (ether), 7, 17
Quirinale, Palazzo del, 25

R

rehabilitation, 60, 63
relativity, 40, 41
Riccardi, Niccolò, 38, 39
Ricci, Ostilio, 2
Rome, 9, 24-25, 28-29, 31, 35-38, 42-44, 46, 47, 55, 62, 63

S

Sagredo, Giovanni Francesco, 13, 19, 21, 22, 39, 52, 53
Salviati, Filippo, 39-41, 52, 53
San Matteo, convent, 27
Santa Croce, church, 57
Santa Maria Sopra Minerva, convent, 45, 63
Sarpi, Paolo, 13, 15, 21
Saturn, 4, 6, 18, 20, 36, 41, 61, 72
Scheiner, Christoph, 26, 62

scientific method, 32
Scripture (s), Holy, 28, 29, 35, 38, 49, 50
Seghizzi, Michelangelo, 30, 44
Siena, 3, 45, 46, 51, 63
Simplicio, 39, 42, 52, 53
star (s), new, 14, 15, 33, 34, 40, 41, 61, 72
Starry Messenger (Sidereus Nuncius), 17, 19, 24, 61
Stevin, Simon, 71
Sun, 4, 6, 8, 18, 20, 23, 26, 29, 32-35, 38, 40, 41, 49, 56, 73
sunspot (s), 25, 26, 41, 62

T

Tartaglia, Niccolò, 2
Tasso, Torquato, 3, 71
telescope (s), 16, 17, 19, 24-26, 61
tide (s), 29, 34, 42
Torricelli, Evangelista, 56, 59, 63
trial, 43, 44, 49, 54, 60, 62, 63
Two New Sciences, 13, 15, 52, 54, 55, 63, 74

U

universe, 4, 6, 7, 19, 32, 33, 40, 42, 48, 59
Urban VIII, see Barberini, Maffeo, Cardinal

V

Vallombrosa, Monastery of, 1
Venetian Republic (the *Serenissima*), 6, 11, 13, 16, 19, 22
Venice, 6, 13, 16, 17, 21, 22, 39, 52, 56, 61
Venus, 4, 6, 20, 23, 25, 33, 41, 61, 72
phases of, 25, 40 41, 61
Villa delle Selve (Signa), 39
Villa Il Gioiello (Arcetri), 58
Villa Medici (Rome) 29, 43, 45, 63
Villa Segni (Bellosguardo), 31
Viviani, Vincenzo, 3, 5, 56, 58, 61, 63, 71, 74

------ o ------

Notes

1. *The Age of Pisa*

[1] Galileo was the last of the great Italians, such as Dante, Leonardo, and Michelangelo, to be called by his first name only.

[2] Paolo Scandaletti, *Galileo Privato*, Gaspari, 2009, p. 55 p. 121, pp. 127–8, p. 84, pp. 88–9, p. 150.

[3] Vincenzo Viviani, *Racconto istorico della vita di Galileo*, in A. Favaro, *Le Opere di Galileo Galilei*, vol. XIX, Barbera, 1968.

[4] Many historians doubt that Galileo's discovery of the isochronism of the pendulum took place in Pisa in those years. They think that perhaps it took place during the first years of his stay in Padua, when he made his most remarkable discoveries about the pendulum.

[5] Ludovico Ariosto (1474–1533) was Galileo's favorite poet; he wrote *Orlando Furioso* (*The Frenzy of Orlando*). Torquato Tasso (1544–1595) wrote *Gerusalemme Liberata* (*Jerusalem Delivered*). The three parts of the *Divina Commedia* (*Divine Comedy*), written by Dante Alighieri (1265–1321), are titled: *Inferno, Purgatorio, Paradiso*.

[6] Historians also doubt the story about Galileo and the Leaning Tower of Pisa. However, he would not have been the first to run this kind of demonstration: Simon Stevin in the Netherlands had done the same in 1586.

[7] Stillman Drake, *Galileo Galilei*, in *Dictionary of Scientific Biography*, Charles Coulton Gillispie, 1972, pp. 247–8.

[8] Galileo Galilei, *Against the Donning of the Gown* (available in Italian and English translation in: Giovanni F. Bignami, *Galileo Galilei*, Moon Books Ltd., 2000).

2 *Professor at Padua*

[9] Bernard Faidutti, *Copernic, Kepler & Galilée Face Aux Pouvoirs*, L'Harmattan, 2010, p. 243, p. 276, pp. 328–9.

[10] Maurice A. Finocchiaro (Editor), *The Essential Galileo*, Hackett Publishing Company, Inc., 2008 (eBook).

[11] Jean-Yves Boriand, *Galilée*, Perrin, 2010, pp. 61–2, p. 127.

[12] The new stars of 1572 and 1604 were supernova*e*—that is, massive stars that exploded catastrophically at the end of their life. They resulted in exceptional bright objects that emitted vast amounts of energy. The new star of 1572 is now called "Tycho's supernova," while that of 1604 is called "Kepler's supernova."

[13] Erminia Ardissino, *Galileo Galilei, Lettere*, Carocci, 2008, pp. 68–9, p. 81, p. 127, p. 198.

[14] The anagram, written in Latin, read: "*Salve umbistineum geminatum Martia proles.*" In November 1610, Galileo finally revealed its meaning, which turned out to be: "*Altissimum planetam tergeminum observavi.*" Galileo's telescopes were not powerful enough to see the details of the curious appearance of Saturn. Nearly fifty years later, the Dutch astronomer and mathematician Christiaan Huygens discovered that Saturn was surrounded by a ring.

[15] Antonio Favaro, *Galileo astrologo secondo i documenti editi e inediti*, Mente e Cuore (a periodical), 1881, pp. 1–10.

3 *Return to Florence*

[16] Galileo's anagram read: "*Haec immatura a me jam frustra leguntur o y,*" which turned out to be: "*Cyntiae figures aemulatur mater amorum*" ("The mother of love [Venus] emulates the figures of Cynthia [the Moon]"). It means: "Venus imitates the phases of the Moon."

[17] Ludovico Geymonat, *Galileo Galilei*, Einaudi 1969, p. 57, p. 72, p. 187, p. 154–5.

[18] Arthur Koestler, the author of *The Sleepwalkers*, writes: "After his sensational discoveries of 1610, Galileo will abandon research as the astronomical theory to devote himself to his propaganda crusade."

[19] Michael Sharratt, *Galileo, Decisive Innovator*, Cambridge University Press, 1994, p. 140.

[20] Annibale Fantoli, *Galileo e la Chiesa*, Carocci, 2010, p. 89.

[21] James Reston, *Galileo: A Life*, Beard Books, 2000, p. 189.

4 *The Dialogue and the Abjuration*

[22] Galileo, *Dialogue Concerning the Two Chief World Systems*, transl. Stillman Drake, University of California Press, 1967.

[23] Galileo always claimed never to have read *Astronomia Nova*, the revolutionary book that Kepler published in 1609. Historians are divided on this issue: many doubt the sincerity of Galileo. Ludovico Geymonat, who was very supportive of Galileo, acknowledged that "it is difficult to explain that behavior." Albert Einstein said: "I was always shocked that Galileo had not acknowledged the work of Kepler."

[24] About fifty years after Galileo's death, Isaac Newton discovered the universal law of gravity, governing planetary motions. Consequently, Earth and other planets have to go around the sun, because of its much larger mass. In 1625, James Bradley discovered "stellar aberration" (slight change in stellar positions due to Earth's speed), which was a proof of the earth's orbital motion. Another consequence of this motion was "stellar parallax" (slight change in stellar positions due to Earth's changing position), first measured by Friedrich Bessel in 1838. Another discovery proving Earth's orbital motion was that of the Doppler effect (slight change in color of stars due to Earth's speed), discovered by Christian Doppler in the first half of the

nineteenth century. In 1851, Léon Foucault performed an experiment with a pendulum, which demonstrated for the first time the diurnal rotation of the earth.

[25] Urban VIII had elevated two of his nephews and his brother to the position of cardinal. He was also criticized for having built (for his nephew Cardinal Francesco) *Palazzo Barberini* with the marble of the ancient Rome monuments. Among the people, the following lampoon circulated: "*Quod non fecerunt Barbari, fecerunt Barberini*" ("What the barbarians did not do, the Barberini did").

[26] Stillman Drake, *Galileo, A Very Short Introduction*, Oxford University Press, 2001 (eBook).

[27] Dava Sobel, *Galileo's Daughter*, Harper Collins eBooks, 2010.

5 *Years of Solitude*

[28] Galileo Galilei, *Two New Sciences*, transl. in English by Stillman Drake, Wall & Emerson, 1989, p. 147, p. 148, p. 149.

[29] Vincenzo Viviani wrote [3]: "Such was the delight he had in the delicacy of wines and grapes that he himself, with his own hands, pruned and tied the grapevines, in the gardens of his villas, with exceptional diligence."

[30] http://www.fisicamente.net/FISICA_2/index-1855.pdf

[31] See Annibale Fantoli, *Galileo e la Chiesa*, Carocci, 2010, pp. 209–17. See Michael Sharratt, *Galileo*, Cambridge University Press, 1996, pp. 209–22.

www.ingramcontent.com/pod-product-compliance
Lightning Source LLC
Chambersburg PA
CBHW071251170526
45165CB00003B/1301